Lyamine BRIKI

Soil mechanics for beginners

Lyamine BRIKI

Soil mechanics for beginners

ScienciaScripts

Imprint

Any brand names and product names mentioned in this book are subject to trademark, brand or patent protection and are trademarks or registered trademarks of their respective holders. The use of brand names, product names, common names, trade names, product descriptions etc. even without a particular marking in this work is in no way to be construed to mean that such names may be regarded as unrestricted in respect of trademark and brand protection legislation and could thus be used by anyone.

Cover image: www.ingimage.com

This book is a translation from the original published under ISBN 978-620-6-70490-4.

Publisher:
Sciencia Scripts
is a trademark of
Dodo Books Indian Ocean Ltd. and OmniScriptum S.R.L publishing group

120 High Road, East Finchley, London, N2 9ED, United Kingdom
Str. Armeneasca 28/1, office 1, Chisinau MD-2012, Republic of Moldova, Europe
Printed at: see last page
ISBN: 978-620-7-72460-4

Copyright © Lyamine BRIKI
Copyright © 2024 Dodo Books Indian Ocean Ltd. and OmniScriptum S.R.L publishing group

Contents

Chapter I

Introduction to soil mechanics

Chapter 1

Introduction to soil mechanics

1.1 Purpose of soil mechanics

The structures use the ground as much as an infrastructure element that transmits the overall load of the structure to a sufficiently stable and resistant layer of ground. As a result, the success of the structure depends on the success of the foundation project. Depending on the type of structure and the way it is designed, the soil may form a support base for the entire structure, such as a road, tunnel, gravity dam, retaining wall, aërodrome, or a support point for only a few ëlëments, such as a building, bridge, arch dam, etc. Soil mechanics is the science that brings together all the knowledge and techniques that enable :

- Identify the characteristics that govern the mechanical behaviour of the soil.
- Analysis of soil-structure interaction
- Correct installation of underground structures.

By way of example, soil mechanics deals with problems relating to various foundations, support structures, embankments and earth structures, slope and embankment stability, roads, airstrips, tunnels, mines, etc.

1.2 Soil mechanics disciplines :

In order to achieve the above objectives, several disciplines will be necessary:

a- Geology of the terrain : The study of the geology of the terrain is of great importance. In fact, it makes it possible to identify the different soil layers, their thickness and dip, as well as the possible presence of groundwater. In addition, the geological study of the layers present gives qualitative descriptions of the soil, answers some questions relating to the history of the dëpдt and helps to orient preliminary research.

b- Physico-chemical characteristics: The study of the physical and chemical characteristics of soils has shown that they are very useful in predicting or interpreting soil behaviour. Most of these properties are determined by laboratory or on-site tests.

c- Hydraulic study: The presence of water in the various layers plays a major role in the behaviour of the soil. Determining the level of stabilisation and studying the flow regime enables the choice of pumping and dewatering equipment, as well as counteracting the phënomënes of boiling sands. Determining the chemical nature of the groundwater makes it possible to predict how buried structures will be ëtanchëitë.

d- Mechanical characteristics: Analysis of the mechanical behaviour of soils is based on the conclusions of the preceding disciplines and on laboratory or site tests. This discipline makes it possible to determine the resistance of the soil and its bearing capacity, and therefore the choice of foundation method and the dimensions of the buried ëlëments. Lastly, it makes it possible to predict quantitatively the dëformation or settlement of the ground under the load of the structure.

e- Theoretical research and numerical modelling: Several theories have been developed **with the** aim of understanding complex physical phenomena. They describe the problems posed by rigorous mathematical models, the resolution of which makes use of increasingly advanced computer and numerical techniques, and occupy a large part of current research in this field.

f- Design and implementation: These are the techniques used to design and build underground structures. It takes into account the study of the costs of the various possible solutions.

In addition to know-how, the regulations in force must be followed step by step to guarantee safety conditions both during construction and during operation of the structure.

1.3 Aspect of the earth's crust :

1.3.1 Constitution of the Earth :

- An envelope, the earth's crust or lithosphere (thickness ~ 70 km, composition: largely silica and aluminium, density ~ 2.7),
- An intermediate, pasty layer, the pyrosphere (~*3000* km thick), composition: rich in silicon and magnesium, density ~ 3.4),
- A central core, or barysphere (radius ~ 3300 Km, composition: mainly nickel and iron, density ~ 7.8).

The cooling of the Earth has caused dislocations and contractions, creating emergences, folds and faults that have evolved over time (the age of the Earth is estimated at 4 billion years).

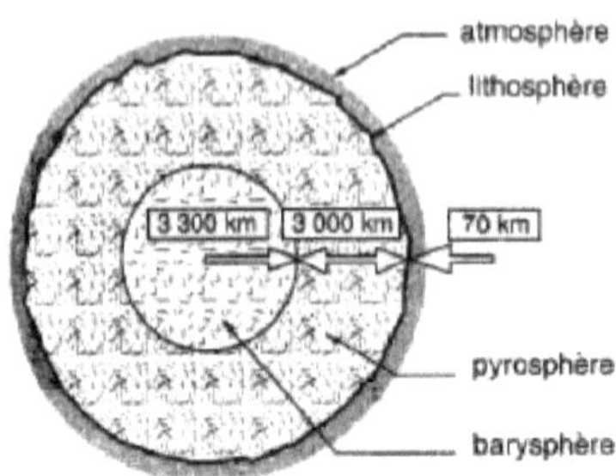

Torn from the primitive rocks by rain, hail, snow, ice, wind, alternating frost and thaw, waves and seas, then transported into the hollows of the folds of the crust, the sëdiments gave rise to the sedimentary rocks.

They are the result of chemical (oxidation), physical (variation in tempërature, frost) or mechanical (erosion, waves, etc.) alteration of the rocks.

This transformation of loose, superimposed rock (mud and sand) into cohesive rock takes millions of years.

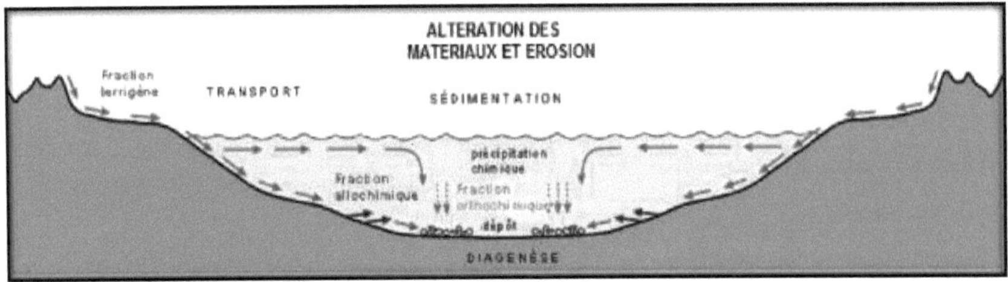

Sedimentary rocks are therefore composed of alternating beds, banks or strata (layers) bounded by visible or invisible joints. These beds are not continuous; they end either by losing their strength (thickness) or by modifying the facies (set of metallogical and structural characteristics) of the rock.

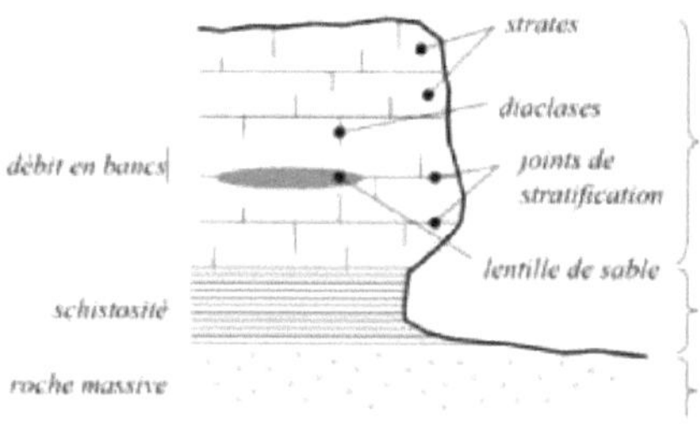

Most rocks are c^1a1^ëe3, cohërent and not very dëformable (quartz, granite, sandstone, etc.). Some are loose (sands) or plastic (clays), others are liquid ^oil).

1.3.2 General points and definitions :

Soil mechanics is a young science. The first foundations can be attributed to Coulomb (1773), but Terzaghi vëritably тШё modern soil mechanics (1936).

Soil mechanics focuses on the behaviour of loose, compressible soils (excluding the behaviour of rocks), in terms of their resistance and dëformability.

Laboratory tests are becoming increasingly sophisticated, based on the analysis of samples at different depths, or on equipment that tests the soil directly at different depths. They provide data on the bearing capacity of the soil, its homogeneity and density.

Soil mechanics provides builders with the data they need to design civil engineering and building structures and ensure their stability in relation to the soils on which they are to be founded:

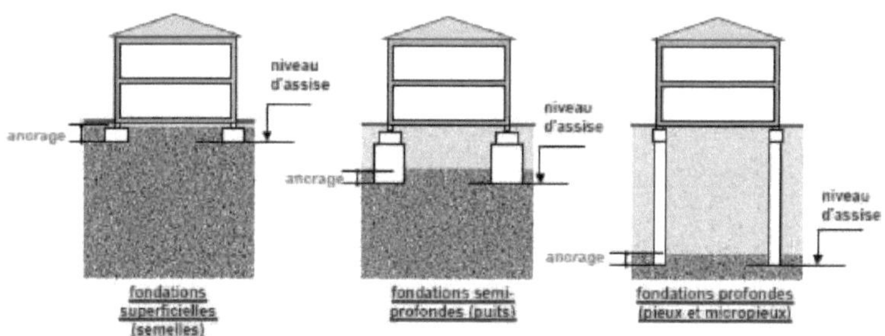

(Foundations operate on the action-reaction principle, transmitting loads into the ground and thus ensuring the stability of the structure. However, the presence of moving water in the ground is a source of structural changes and can pose a risk to stability.)

Or with which they will be built:

✓ Embankments (roads, railways, dams, dykes, earthen basins, maritime platforms, etc.)

✓ Embankments (embankments, canals, basins, etc.)

and secondly, support (retaining walls, sheet piling, etc.)

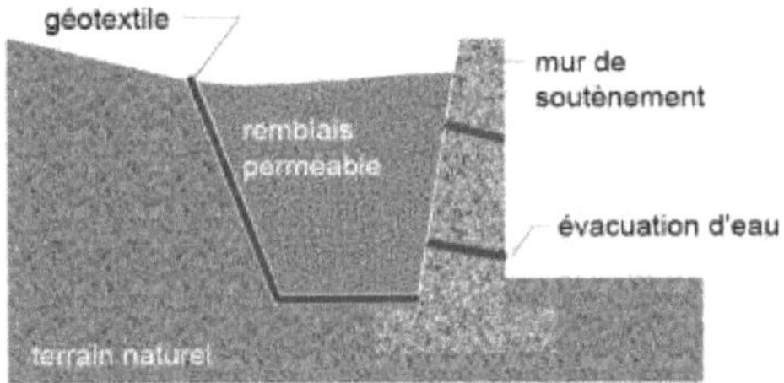

1.4 Dimensional characteristics :

1.4.1 Form :

Three catëgories of shapes can be distinguished:

✓ Sphërical (cubic) particles with rounded or angular shapes in the case of granular soils ;

✓ Platelet particles in fine soils ;

Needle-shaped particles.

1.4.2 Size :

The solid skeleton of a soil is formedë of particles of very varied dimensions, the Swedish Atterberg proposes the following classification:

Pulverulent soils (Grenus)$_L$ (Coherent)

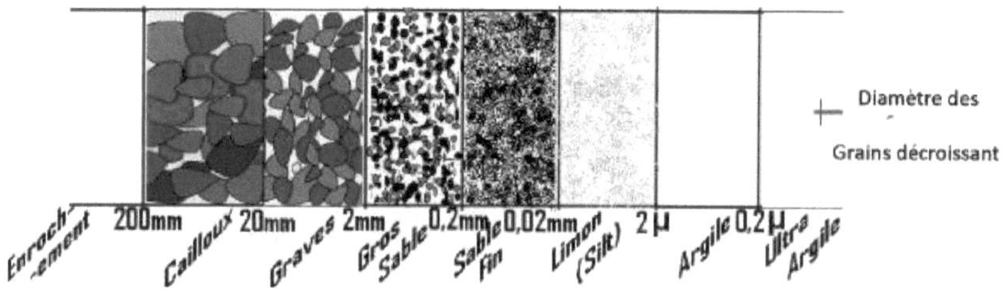

1.5 Granulometric characteristics :

Diffërent techniques for obtaining granulometry according to soil type:

a- Pulverulent soils: sieving is used (a series of "circular hole" sieves or a series of "square mesh" sieves).

- dry for large grains ;

- underwater for fine grains.

b- Fine soils :

We use sëdimentomëtrie or laser granulometry

After removal of carbonates, organic substances and possible iron oxides (because of their binding function), the fine fraction (<50pm), is suspended in tubes filled with water. The method is based on the difference in sëdimentation speed between the lightest and coarsest particles, using Stockes' law.

$$V = \frac{h}{t} = \frac{2\,(dp-d)\,gr^2}{9\eta}$$

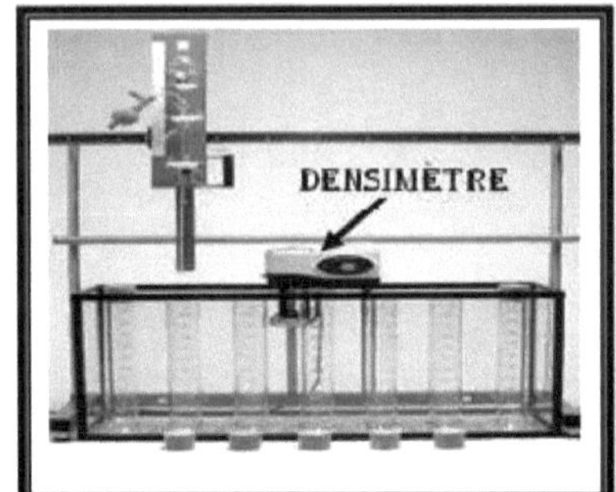

V = sedimentation velocity, cm/sec g = gravitational acceleration, cm/sec^3 dp = particle density, g/cm^3 d = liquid density, g/cm^3 i' = particle radius, cm i] = viscosity site, g/cm/sec

c- Organic soils: certain soils, derived from recent gëological dëpдts, may contain organic matter. They can be identified by their grey to black colour, the presence of vegetable debris and their odour. They have poor technical properties and in particular those containing 2 to 3% organic matter can cause long-term settlement problems when used as backfill.

Peat, the result of the decomposition of plant fibres, is an example of organic soil, and is composed almost exclusively of plant fibres.

Chapter II

Soil identification and classification

Chapter 2

Soil identification and classification

2.1 Components :

Soil is a collection of grains ттёраих and organic matter dëcomposëes (solid particles), sëparable by mëcanic action 1ёдёге hence the name SOIL MECHANICS and liquid and gas occupying the empty spaces between the solid particles.

A soil is therefore a three-phase mixture:

> Solid phase composed of grains (sand, clay, etc.),

> A gaseous phase composed of air, when the soil is dry, or a mixture of air and water vapour, when the soil is damp. When all the voids are filled with water, the soil is said to be saturated,

> Liquid phase composed of water filling the voids between the particles.

> **.1.1 Physical characteristics of soils :**

This involves defining certain parameters relating to the various proportions in which the skeleton, water and gas in a soil are found.

Schematically, we consider the following presentation of a ë soil sample in which the three phases would be sëparëes.

The volumes of the various phases are shown on the right and the weights on the left.

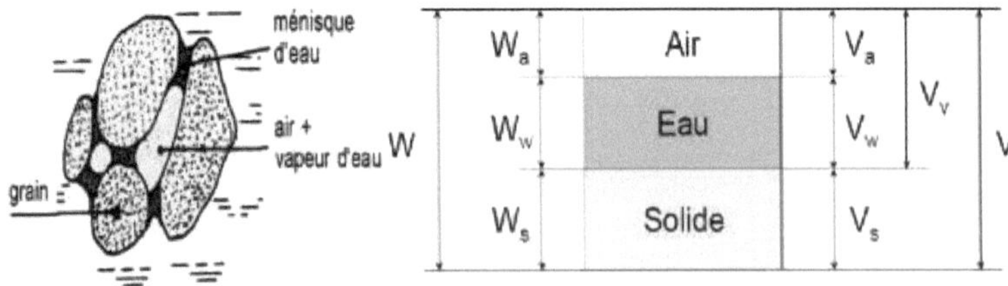

W for weight

w for water

\N I total weight of soil

W_s I weight of solid particles$W = W_s + W_w$

W_w : weight of the water

V: total volume (apparent)

V_c I volume of solid particles . .. . . .

s V = V + V

V": volume of voids between particles . / . .^{w} . /^{a} V = V + V =

V_w : volume of water s v

V": volume of air

LI

2.2 Dimensional parameters :

Density

$$\rho = \frac{M}{V} [Kg/m^3]$$

Density : $\gamma = \rho . g [N/m^3]$

Most questions use :

Density :

Solid grains: $\gamma_s = \frac{w_s}{v_s}$ sand and clay " 26 to 27 kN/m3

The solid phase of soils consists mainly of silica and alumina. Organic soils are an exception to these values.

Water: $\gamma_w = \frac{w_w}{v_w} =$ 9.81kN/m3

Soil (or bulk or wet weight)

$\gamma = \frac{w}{v}$ Sand " 17 to 20kN/m3 ; clay " 16 to 22kN/m3

Dry: $\gamma_d = \frac{W_S}{V}$ sand " 14 to 18kN/m3 ; clay " 10 to 20kN/m3. d for dry

d pour dry

If the soil is dry, then $\gamma = \gamma_d$.

Saturated soil: when all the voids are filled with water:

$\gamma_{sat} = \frac{W}{V} = \frac{W_S + \gamma_w * V_v}{V}$ Sand and clay " 19 to 22kN/m3.

From dëjaugë soil: $\gamma' = \gamma_{sat} - \gamma_w$ sand and clay " 9 to 12kN/m3

2.3 Dimensionless parameters :

They indicate the proportions of the various phases in a soil:

❖We define porosity, notëe n, which is used to determine the size of voids, i.e. whether the soil is in a loose or tight state. It is defined by :

$n = \frac{V_v}{V}$ Sand " 0,25 a 0,50 ; clay " 0,20 a 0,80

The porosity is always less than 1.

It can be expressed as a percentage.

■ The void index, denoted e, is defined by :

$$e = \frac{V_v}{V_s}$$

Sand " 0,5 a 1 ; clay " 0,3 a 1

The index of can be greater than 1.

It can be expressed as a percentage.

❖ The water content, denoted w, is defined as the ratio of the weight of water to the weight of solid particles in a given volume of soil:

$$w = \frac{W_w}{W_s} . 100$$

Sand = 1 to 15%; clay = 10 to 20%.

w can exceed 100%.

❖ The degree of saturation, Sr, indicates the proportion to which the voids are filled with water. It is defined as the ratio of the volume of water to the volume of voids.

$$S_r = \frac{V_w}{V_v} . 100$$

Sr = 100% -→ the soil is saturated; Sr = 0% → the soil is dry

Of all the parameters defined above, the dimensionless parameters are the most important. They characterise the state of the soil, i.e. the state of compaction of the skeleton and the quantities of water and air contained in the soil.

We also introduce the notion of density in relation to water

Wet density : $d_h = \frac{\gamma_h}{\gamma_w}$;

Dry density : $d_d = \frac{\gamma_d}{\gamma_w}$

❖ Grain density[1] : $d_s = \frac{\gamma_s}{\gamma_w}$;

❖ Tidal density : $d_j = \frac{\gamma'}{\gamma_w}$

2.4 Practical determination of physical characteristics :

In view of the inevitable dispersion of the results, a large number of measurements of three independent parameters are taken and the average is taken.

2.4.1 Determination of water content (ponderale) w:

A first weighing of the sample in its initial state gives the mass **m** of the wet sample and a second weighing after oven drying at 105°C for 24 hours (evaporation of free water and capillary water), gives the dry mass of the sample **ms**.

Weighing and drying at 105 C

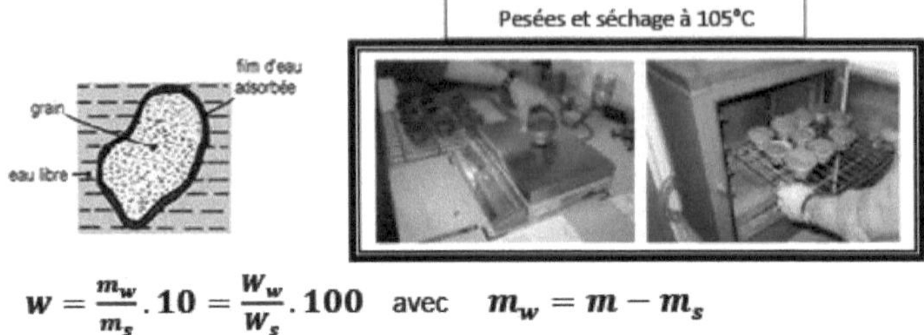

$$w = \frac{m_w}{m_s}.10 = \frac{W_w}{W_s}.100 \quad \text{avec} \quad m_w = m - m_s$$

2.4.2 Determination of density y :

The mass m and total volume V of the sample must be determined. To determine the latter, one of the following three methods is used:

- **Water immersion method :**

A natural sample in a simple, unworked form, with a mass of between 0.1 and 0.5 kg, is weighed (m) and then covered with a layer of paraffin $(\rho_{paraffine} =$ 0.88 *g/cm?).* A second weighing (mp) is used to determine the mass of the paraffin layer and calculate its volume. A third, hydrostatic weighing of the paraffin-coated sample (m'p) is used to calculate the volume of the paraffin-coated sample. As the volume of paraffin is known, the volume V of the sample is deduced:

$$V = V_{eau+paraffine} - V_{paraffine} = \frac{m_p - m'_p}{\rho_w} - \frac{m_p - m}{\rho_p}$$

- **Cutting kit method :**

The sample of soil in its natural state is punched with a cutting tool. The faces of the test sample are flush at the extremes. The volume V of the test sample is equal to the product of the cross-sectional area of the cutting tool and its height.

It should be pointed out that the soil sample is heavily modified by the passage of the cutting tool.

- **Mould method :**

The sample, prepared according to a defined process, fills the mould until it overflows.

The upper end of the mould, of known dimensions, is flush with the rule.

2.4.3 Determination of the density of solid particles$_s$:

A known mass m_s of sëchë soil (by oven drying at 105°C to constant mass) is introduced into a container containing distilled water. A magnetic stirrer separates the particles from each other. The released air bubbles are sucked up by a vacuum (water tube). After ensuring that no air bubbles are trapped between the solid particles, the volume of water displaced by the solid particles is determined with great care.

The volume of the solid phase Vs, equal to the volume of water displaced by the soil, is determined by weighing.

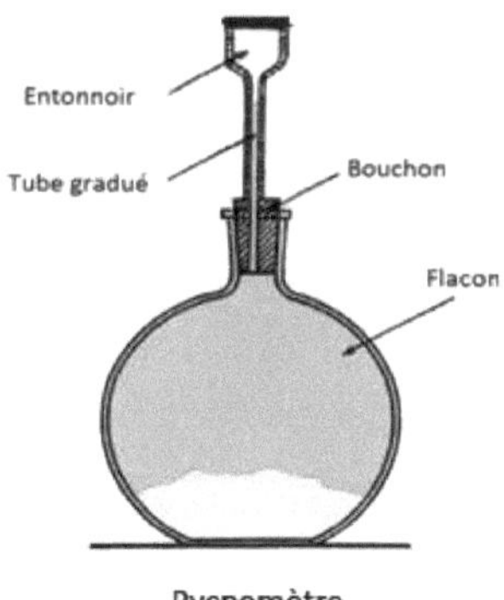

Pycnomètre

m_1: mass of pycnometer containing distilled water and magnetic rod, m_2: mass of pycnometer containing soil, distilled water and magnetic rod.

$$m_2 = m_1 + m_s - \rho_w V_s$$

With m_s: mass of solid particles, ρ_w: density of distilled water, Vs: volume of solid particles.

$$V_s = \frac{m_1 + m_s - m_2}{\rho_w}$$

$$\rho_s = \frac{m_s}{V_s} = \frac{m_s}{m_1 + m_s - m_2}\rho_w \Rightarrow \gamma_s = \rho_s \cdot g$$

Application 1:

A ë sample of saturated clay has a mass of 1.526kg; after oven drying its mass is only 1.053kg. The solid constituent of the grains has a density^ of 2.7. We ask :

- Water content w;
- The videse index ;
- The poros^ n;
- The wet density yh;
- Wet density (yh/yw)

We'll take g = 9.81 m/s^2

Solution 1:

$$w = \frac{W_w}{W_s} = \frac{M_w}{M_s} = \frac{1.526 - 1.053}{1.053} = \frac{0.473}{1.053} = 0.449$$

→w=45%

$$e = \frac{V_v}{V_s} = \frac{V_w}{V_s} = \frac{M_w}{\rho_w} \cdot \frac{\rho_s}{M_s} = w\, d_s = 45\% * 2.7 = 1.215$$

$$n = \frac{V_v}{V} = \frac{V_w}{V} = \frac{V_w}{V_s + V_w} = \frac{M_w}{\rho_w} \cdot \frac{1}{\frac{M_w}{\rho_w} + \frac{M_s}{\rho_s}} = \frac{M_w}{M_w + \frac{M_s}{d_s}} = 0.548$$

→n=0.55

$$\gamma_h = \frac{W}{V} = \frac{M * g}{V_s + V_w} = \frac{M * g}{\frac{M_s}{\rho_s} + \frac{M_w}{\rho_w}} = \frac{M * g * \rho_w}{\frac{M_s}{d_s} + M_w} = 17.34$$

$$d_h = \frac{\gamma_h}{\gamma_w}; \quad \gamma_w = \frac{W_w}{V_w} = \frac{M_w * g}{V_w} = \rho_w . g = 9.81 KN / m^3$$

→d_h=1.77

Application 2:

One ёсИап^Поп of soil has a mass of 129.1g and a volume of 56.4cm^3 . The mass of the grains is 121.5g. The solid constituent of the grains has a dens^ of 2.7.

We ask:

- Water content ;
- The voids index ;
- The degree of saturation Sr.

We'll take g = 9.81 m/s^2 .

Solution 2:

$$w = \frac{W_w}{W_s} = \frac{M_w}{M_s} = \frac{129.1 - 121.5}{121.5} = \frac{7.6}{121.5} = 0.0625$$

→w=6.3%

$$e = \frac{V_v}{V_s} = \frac{V - V_s}{V_s} = \frac{V}{V_s} - 1 = \frac{V}{\frac{M_s}{\rho_s} \cdot \frac{\rho_w}{\rho_w}} - 1 = \frac{V}{\frac{M_s}{d_s * \rho_w}} - 1 = 0.25$$

$$S_r = \frac{V_w}{V_v} = \frac{V_w}{V - V_s} = \frac{1}{\frac{V}{V_w} - \frac{V_s}{V_w}} = \frac{1}{\frac{V * \rho_w}{M_w} - \frac{M_s}{d_s * M_w}} = 0.666$$

→Sr=67%

Application 3:

A soil with a dry density $d_d = 1.6$, a solid density $= 2.65$ and a water content w = 0.10 (10%).

Calculate :

- Saturated density p_{sat} ,
- Porosite n,
- The void index e,
- The degree of saturation S_r ,
- Water content at saturation w_{sat}.

Solution 3:

$$\rho_s = \frac{M_s}{V_s}; \quad \rho_d = \frac{M_s}{V_t}$$

$$\rho_{sat} = \frac{M_t}{V_t} = \frac{M_s}{V_t} + \frac{M_W}{V_t} * \frac{M_S}{M_S} = \frac{M_S}{V_t}\left(1 + \frac{M_W}{M_S}\right) \rightarrow \rho_{sat} = \rho_d(1 + w)$$

A.N: $\rho_{sat} = 1{,}6(1 + 0{,}1) \rightarrow \boxed{\rho_{sat} = 1{,}76\, {}^{t}/_{m^3}}$

$$\rho_s = \frac{M_s}{V_s} * \frac{V_t}{V_t} = \rho_d * \frac{V_t}{V_s} = \rho_d\left(\frac{V_s + V_v}{V_s}\right) = \rho_d(1 + e)$$

A.N: $e = \frac{\rho_s}{\rho_d} - 1 = \frac{2{,}65}{1{,}6} - 1 = 0{,}65 \rightarrow \boxed{e = 65\%}$

$$S_r = \frac{V_w}{V_v} = \frac{V_w}{eV_s} * \frac{M_s}{M_s} * \frac{M_w}{M_w} = \frac{\rho_s}{e\rho_w} * w$$

A.N: $S_r = \frac{2{,}65}{0{,}65*1} * 0{,}10 = 0{,}40 \rightarrow \boxed{S_r = 40\%}$

The water content at saturation w_{sat}. corresponds to $S_r = 1 \rightarrow 1 = \frac{\rho_s}{e\rho_w} * w_{sat}$

A.N: $w_{sat} = \frac{e*\rho_w}{\rho_s} = \frac{0{,}65*1}{2{,}65} = 0{,}24 \rightarrow \boxed{w_{sat} = 24\%}$ Granulometric curve :

Rejection on a sieve: the quantity of material retained on the sieve.

Sieve: the quantity of material passing through the sieve.

All you have to do is plot the various sieve percentages on a semi-logarithmic sheet:

- Abscissa: the dimensions of the mesh, logarithmic scale;
- On the ordinate: percentages on an arithmetic scale;
- The curve must be continuous.

Semi-logarithmic coordinates provide a more accurate representation of fine particles, which have a major influence on soil behaviour.

2.5 Characterisation of the granulometry of a soil :

The granulometry of a soil can be сaraйёnзёe by l'utilisation de coefficients :

1) Uniformity coefficient or Hazen coefficient :

$$C_u = \frac{D_{60}}{D_{10}} \quad \begin{cases} \text{Pour } C_u > 2 \rightarrow \text{la granulométrie est dite étalée.} \\ \text{Pour } C_u < 2 \rightarrow \text{la granulométrie est dite uniforme ou serrée.} \end{cases}$$

For C_u **> 2 ^ the** particle size is said to be calibrated.

For C_u **< 2 ^ the** particle size is said to be uniform or tight.

DY: sieve opening through which y% of the grain weight can pass.

D10: is called the effective diameter.

2) Coefficient of curvature : $C_c = \frac{D_{30}^2}{D_{60} * D_{10}}$

When certain conditions on C_u and C_c are satisfied, the soil is said to be well graded, i.e. the granulometry is well distributed, without the predominance of any particular fraction.

When the granulometry is discontinuous, with a predominance of one particular fraction, it is said to be poorly graded.

Well-graded soils constitute naturally dense dëpдts with a ëlevëe bearing capacity. They can be easily compacted into fill and form stable slopes.

Application:

3500g of previously sieved sand is sieved dry. It is first observed that the 12.5mm sieve (i) does not retain any fraction of the material. We then use a column of six sieves with the following mesh openings, from top to bottom: 5; 2; 1; 0.5; 0.2; 0.1mm. The rejects on each of the six sieves are as follows (from top to bottom):217g;868g;1095g;809g;444g;39g and the sieve from the last sieve collected on the bottom which closes the column is 28g.

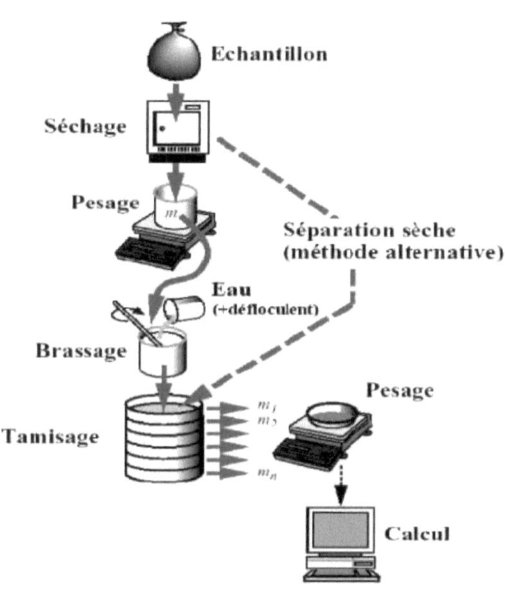

We ask you to construct the granulomëtric curve and to determine the effective diameter

(D_{10}) and the uniformity coefficient^.

Solution:

	Opening strainers	Rejects(g)	Sieve cake (g)	Sieve cake
1	12,5	0	3500	100
2	5	217	3283	94
3	2	868	2415	69
4	1	1095	1320	37,7
5	0,5	809	511	14,6
6	0,2	444	67	1,92
7	0,1	39	28	0,80

Reste		28	0	

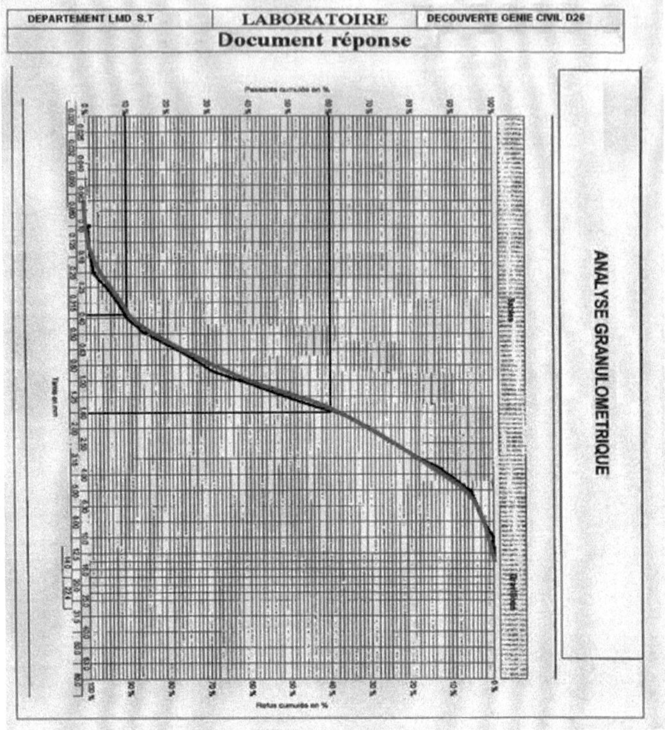

$$C_u = \frac{D_{60}}{D_{10}} = \frac{1{,}60mm}{0{,}37mm} = 4{,}3 > 2 \rightarrow$$ The grain size is spread out.

2.6 Soil structure :

1) Pulverulent (gravelly) soils $D > 20\mu m$ **:**

The grains fall apart under their own weight.

The main forces involved in the equilibrium of the structure are the forces of gravity; it is through grain-to-grain contact reactions that a stable whole can exist. This stabilitë will be all the better if the number of contacts is ë1вyë (well graduated soilë).

2) Fine (coherent) soils $D < 20\mu m$:

The particles remain stuck together. The soil has the appearance of a solid and does not dësagrë under the effect of gravity or other applied forces. The particles are formed by a stack of sheets. They are shaped like platelets.

3) Organic soils :

When the grains are made up of organic matter, the soil is said to be organic. The presence of organic matter in soils, which gives rise to loose textures and high water retention, makes them highly plastic^ and compressible^.

2.7.1 Identification tests specific to granular soils :

2.7.1.1 Sand equivalent test :

The purpose of the sand equivalent test, dësignë by the symbol E.S., is to evaluate the relative proportion of fine elements contained in the soil, the presence of which in significant quantities can modify the mechanical behaviour.

The test consists of performing an energetic wash on the soil sample (fraction of the material whose components are less than 5mm) in order to separate it from its fine matter. The test tube containing the soil and the washing solution is subjected to 90 cycles of 20cm amplitude over 30 seconds. The solution used also has a flocculating effect on the fine particles. It is widely ^тБё, particularly in road gëotechnics.

The solution is allowed to settle. The true sand settles in the bottom of the burette to a measurable level h_2 . Above the sand, the flocculate swollen by the solution settles. A second level h1 can be distinguished which separates the liquid containing the flocculate from the transparent liquid of the dëcantë washing solution. The ratio between the height of the solid dёрдt h_2 and the height of the upper level of the flocculate h1 is determined.

.

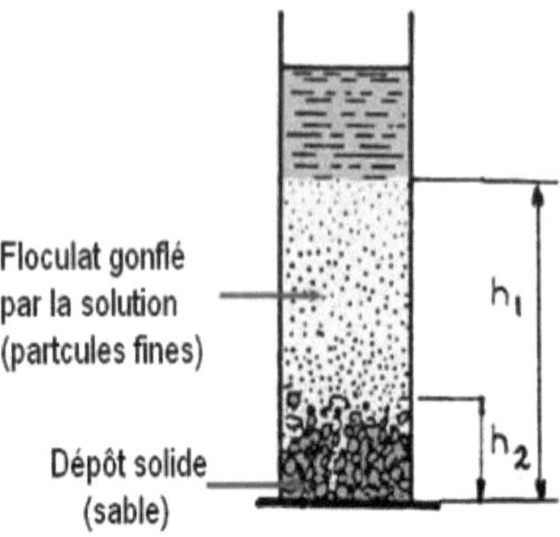

The sand equivalent is by definition :

$$E.S. = \frac{h_2}{h_1} * 100$$

For pure, clean sand E.S.=i00

***2.7.1.2* Density index :**

To give an idea of the state of compactness in which a gritty soil is in its natural state, we define the density index :

$$I_D = \frac{e_{max} - e}{e_{max} - e_{min}}$$

For loose soil $e \approx e_{max} \to I_D \approx 0$

For a greenhouse floor $e \approx e_{min} \to I_D \approx 1$

2.7.2 Identification tests specific to fine soils :

2.7.2.1 Atterberg limits :

The particularity of fine soils is that their consistency varies greatly according to their water content. Their ëtat ranges from solid if they are desséchës, to liquid if they are dëtrempës. Between these two states, there is an intermediary state called plastic (modelling clay).

Atterberg limits are conventional physical constants (weightëral water content) that mark the thresholds between :

- The transition of a soil from a liquid state to a plastic state (liquid limit W)L
- The transition of a soil from a plastic state to a solid state (plastic limit W)p

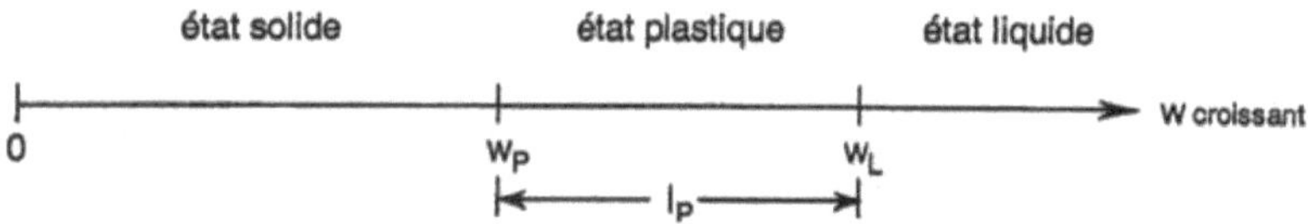

2.7.2.2 Plasticity index IP :

The plasticity index, notë I_p , is the parameter most commonly и1Шзё used to cara^riser the argillosity of soils.

It is expressed by the relationship : $I_P = W_L - W_P$

2.7.2.3 Consistency index IC :

The comparison between the natural soil water content Wnat and the Atterberg limits, provides an idëe of the condition of the soil in place.

To do this, we define the consistency index : $I_C = \frac{W_L - W_{nat}}{I_P}$

liquid very my soft semi-consistent very consistent

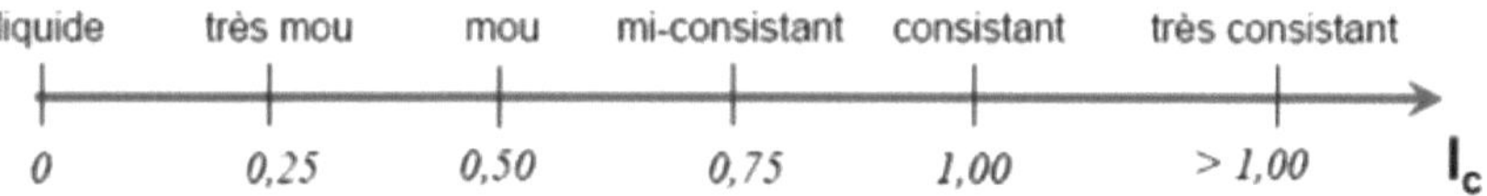

2.8 Soil classification :

Classifying a soil consists of identifying it using quantitative measurements and giving it a name in order to link it to a group of soils with similar characteristics.

Similar to the American U.S.C.S. (United Soil Classification System) classification, the classification used in France by the Ponts et Chaussëes (L.P.C) laboratories is based essentially on granulometric analysis and the plasticity characteristics of the fine fraction, supplemented by very simple tests (colour, odour, effects of water, etc.).

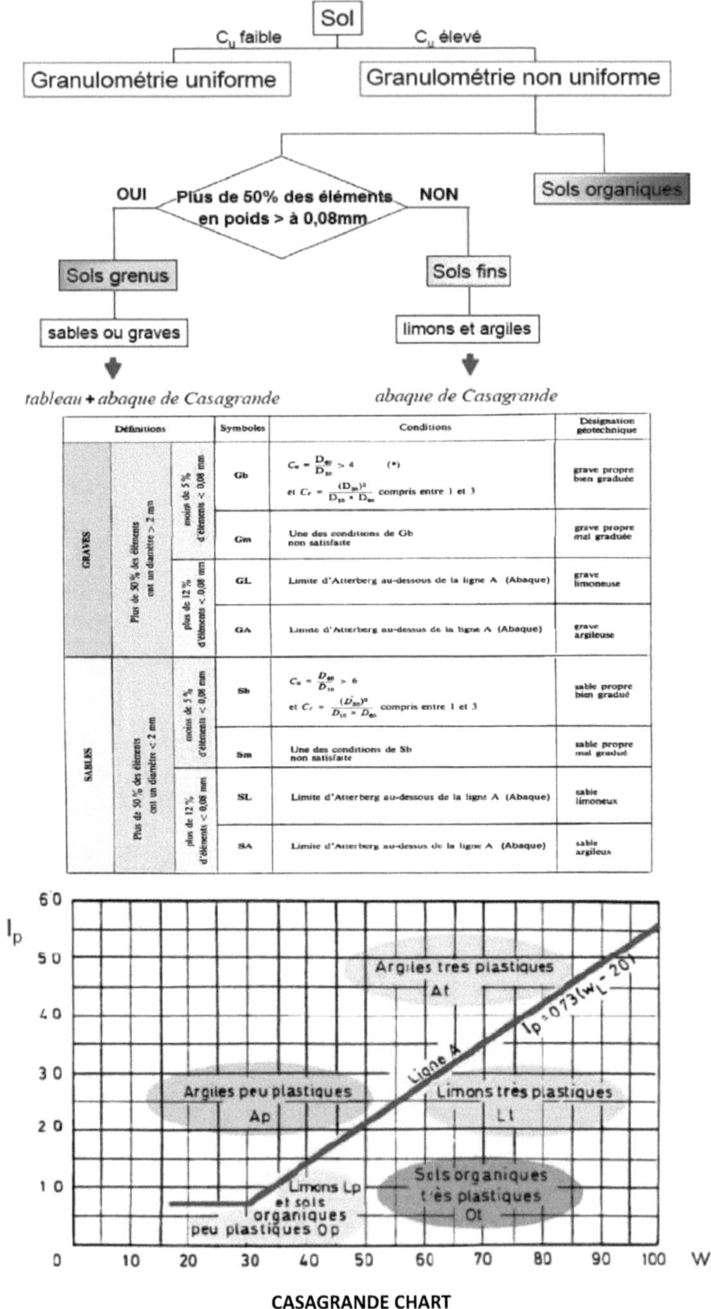

Définitions			Symboles	Conditions	Désignation géotechnique
GRAVES	Plus de 50 % des éléments ont un diamètre > 2 mm	moins de 5 % d'éléments < 0,08 mm	Gb	$C_u = \frac{D_{60}}{D_{10}} > 4$ (*) et $C_c = \frac{(D_{30})^2}{D_{10} \times D_{60}}$ compris entre 1 et 3	grave propre bien graduée
			Gm	Une des conditions de Gb non satisfaite	grave propre mal graduée
		plus de 12 % d'éléments < 0,08 mm	GL	Limite d'Atterberg au-dessous de la ligne A (Abaque)	grave limoneuse
			GA	Limite d'Atterberg au-dessus de la ligne A (Abaque)	grave argileuse
SABLES	Plus de 50 % des éléments ont un diamètre < 2 mm	moins de 5 % d'éléments < 0,08 mm	Sb	$C_u = \frac{D_{60}}{D_{10}} > 6$ et $C_c = \frac{(D_{30})^2}{D_{10} \times D_{60}}$ compris entre 1 et 3	sable propre bien gradué
			Sm	Une des conditions de Sb non satisfaite	sable propre mal gradué
		plus de 12 % d'éléments < 0,08 mm	SL	Limite d'Atterberg au-dessous de la ligne A (Abaque)	sable limoneux
			SA	Limite d'Atterberg au-dessus de la ligne A (Abaque)	sable argileux

CASAGRANDE CHART

Application:

We took 5 soil samples A B CD E and performed the usual identification operations (sieving and limits) on them. The table shows the results:

Sample	Boundaries of Altenberg		Sieve cake		
	W_L	W_P	2mm	0.4mm	80p.m
A	-	-	97	59	0,1
B	23	17	99	93	77
C	28	17	99	76	47
D	-	-	44	8	1
E	23	16	100	85	28
F	28	18	49	33	26

We ask you to apply the L.P.C. classification to these ë samples.

Sample A: More than 50% of ëlëments by weight greater than 80pm (=0.1%) ➔ Gritty soil ;

More than 50% (97%) of the ëlëments have a diameter < 2mm Sand ➔

Less than 5% d^^ments < a 0.08mm (0.1%) ➔ Clean sand well д^иё Sb if the conditions on Cu and Cc are met otherwise it is clean sand poorly д^иё Sm.

Sample B: Less than 50% of ëlëments over 80pm (77%) ➔ Sol fin

Atterberg limit ➔ Slightly plastic clay Ap.

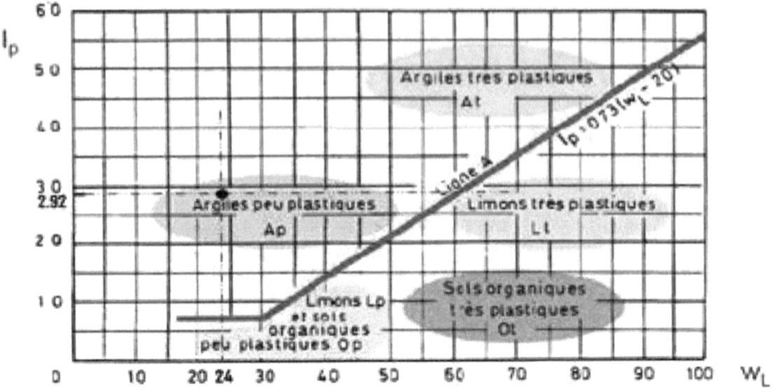

Sample C: More than 50% of the ëlëments above 80pm (47%) ➔ Gritty soil

More than 50% (99%) of the ëlëments are < 2mm in diameter ➔ Sand

More than 12% of the deposits have a diameter of less than 0.08mm (47%) ➔ see the Casagrande diagram:

Limit of Atterberg above the A line ➔ Clayey sand.

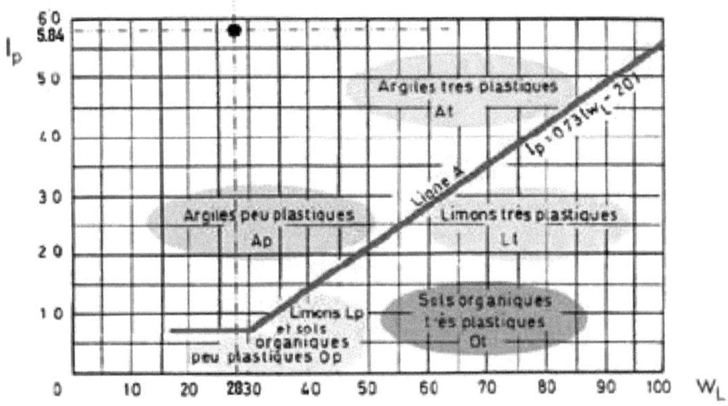

Sample D: More than 50% of ëlëments above 80pm (1%)^ Grey soil

Less than 50% (44%) of ëlëments are < 2mm in diameter ➔ Serious

Less than 5% of elements < a 0.08mm (1%) ➔ Well graded clean gravel Gb if the conditions for Cu and Cc are met, otherwise poorly graded clean gravel Gm.

Sample E: More than 50% of elements greater than 80p.m (28%) ➔ Grey soil

More than 50% (100%) of the elements have a diameter < 2mm Z Sand ➔

More than 12% of elements have a diameter < 0.08mm (28%) ➔ see the Casagrande diagram:

Limit of Atterberg above the A line ➔ Clayey sand.

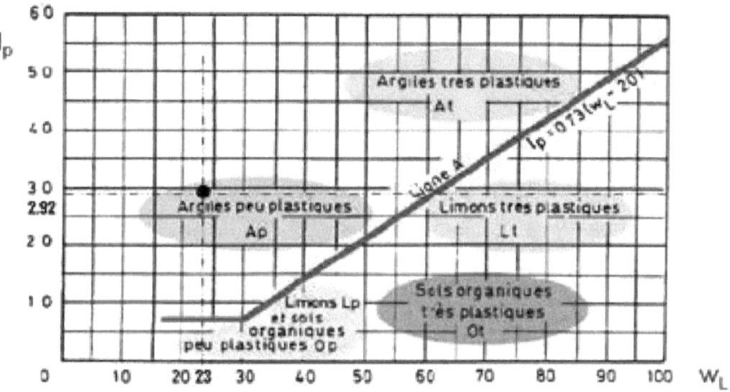

Sample F: More than 50% of the elements greater than 80p.m (26%) ➔ Grey soil

Less than 50% (49%) of the elements have a diameter < 2mm ➔ Serious

More than 12% of elements have a diameter < 0.08mm (26%) ➔ see the Casagrande diagram:

Limit of Atterberg above the line ➔ Gravel Clay.

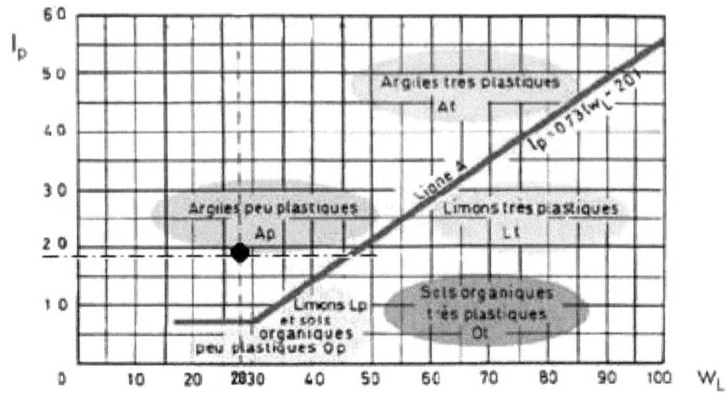

I_p
Argiles très plastiques
At
$I_p = 0.73(w_L - 20)$
Ligne A
Argiles peu plastiques
Ap
Limons très plastiques
Lt
Limons Lp
et sols
organiques
peu plastiques Op
Sols organiques
très plastiques
Ot
W_L

Chapter III

Soil compaction

Chapter 3

Soil compaction

3.1 Definitions - General :

❖ Soil compaction consists of reducing the volume of the soil (or, to put it another way, increasing its density[1]) by applying a mechanical process. This action increases the soil's compactness by tightening its texture. As a result, it reduces the soil's potential for deformation and improves its bearing capacity. It should be noted that there are other methods of increasing the bearing capacity of a soil, such as the addition of binders, but these methods are more expensive.

❖ It is important to realise that the reduction in volume due to compaction only occurs through the elimination of the air-filled voids that exist in the initial soil. The water content w of the soil is not altered!

❖ However, if the water content w of the soil remains unchanged, the same does not apply

for the weight by volume γ_h of the soil which will be increased as well as for the weight

dry volume γd since : $\gamma_d = \frac{\gamma_h}{1+\omega}$

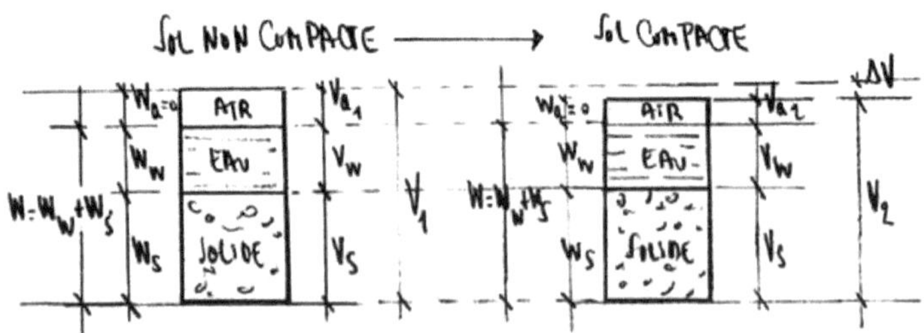

❖ Only V and Va decrease, W ,Ws , Ww ,Vs ,Vw remain unchanged.

❖ As a result, $\gamma_h = \frac{W}{V}$; $\gamma_d = \frac{W_S}{V}$ is on the rise.

❖ The dry weight density y^ is the parameter whose variations due to compaction will be studied. This parameter retains its significance even if the soil under consideration is not dry! In fact, γd is the weight by volume that the soil would have if it were dry.

NB: the higher the , the better the compaction → the aim is to obtain the highest possible γd.

❖ Compaction is influenced by the following factors:

> Water content ;

> The nature of the soil ;

> Compaction energy.

3.2 Influence of water content :

- If a soil is compacted by a series of shocks imparted to it by a falling mass, the dëformation of the soil, i.e. the reduction in its volume, absorbs an energy equal to the potential energy of the mass before the fall (= mgh for a mass m falling from a height h) multiplied by the number of blows given ^ this energy is called the compaction energy.
- If we plot the y^ obtained after compacting a soil sample, by carrying out a series of compaction tests, each corresponding to a different water content (controlled by adding water) but always using the same compaction energy, we obtain the following curve, known as the compaction curve:

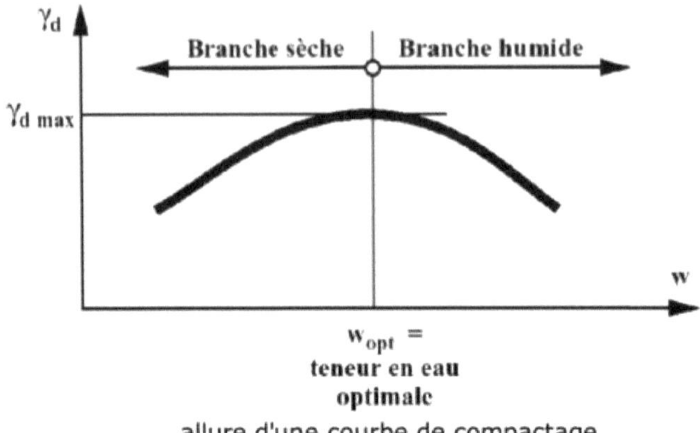

allure d'une courbe de compactage.

optimum water contentc the shape of a compaction curve.

- This is the compaction curve corresponding to a given compaction energy (if the compaction energy is varied, a different curve is obtained).
- The test carried out to establish this curve is known as the Proctor test.
- The curve looks like a bell. It passes through a high point called the optimum

Proctor. is the optimum that a given soil can be made to achieve for a given compaction energy and it corresponds to a well-defined but spëcific water content w ёда1ете^ to the compaction energy implemented.

- The bell-shaped curve can be explained as follows:
- When the water content is ëlevëe, the water absorbs a large part of the compaction energy without any benefit.
- On the other hand, water acts as a lubricant in the compaction process (soil that is too dry causes significant friction between the soil grains).

3.3 Influence of soil type :

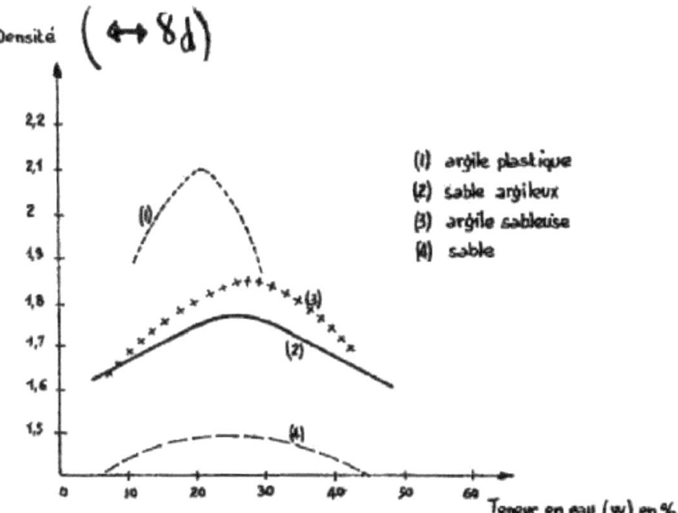

- The shape of the compaction cuts varies with the nature of the soil. They are very flat for sand (which means that γd is only slightly influenced by the water content), and have a very marked maximum for plastic clay (which means that γd is strongly influenced by the water content).

3.4 Influence of compaction energy :

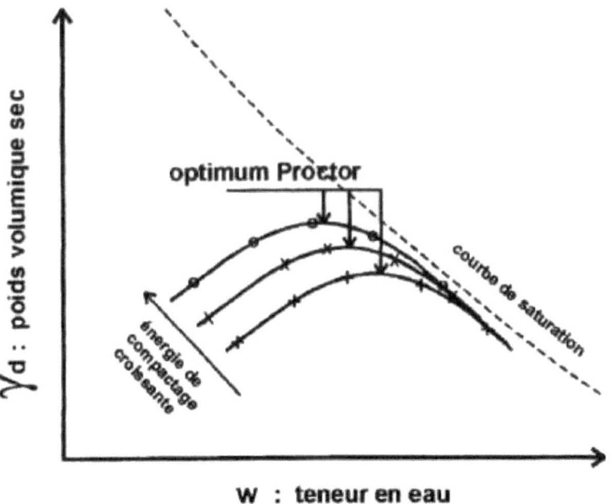

- The compaction curves above correspond to a given soil subjected to differënts compaction Energies $\uparrow$ curves supërieures).

- As compaction energy increases, optimal γd increases and the corresponding water content w decreases.

3.5 Laboratory compaction tests :

- The principle of compaction tests has already been discussed in section 2 (influence of water content). A ësample of soil, of known water content, is compacted into a mould of standard dimensions by the action of a falling tamper. The weight and height at which the lady falls are also standardised. The water content of the soil under test is controlled by drying the sample and then by successive additions of precise quantities of water. A series of tests is carried out for different water contents, the result of each producing a torque (w, γ_d), and therefore a point on the compaction curve.
- The tests can be performed in two types of moulds and in two modes, making four types of test.

Proctor and C.B.R. moulds can be used.

In each of the moulds, the compaction test can be carried out in two ways: the normal Proctor test and the modified Proctor test.

3.6 Practical use of compaction tests :

- Each soil is characterizedërisë by its γ_{dmax} , each of which corresponds to a unique and precise water content for each of the compaction ënergies implemented $\rightarrow$ these values are prëcisëment dëterminë by laboratory tests. Where appropriate, specifications will require soils to be compacted to a γ_d dëterminë.
- The soil on the site has a given water content:

- if this is greater than w_{opt}, we could imagine draining the land, which is practically impossible (A→O) ;
- if this is less than w_{opt}, , consider watering the soil (B→O) ;
- in either case, without altering the water content of the soil, it is possible to increase the compaction energy, i.e. increase the number of passes for a road roller; this last solution is the easiest to put into practice (B→O').

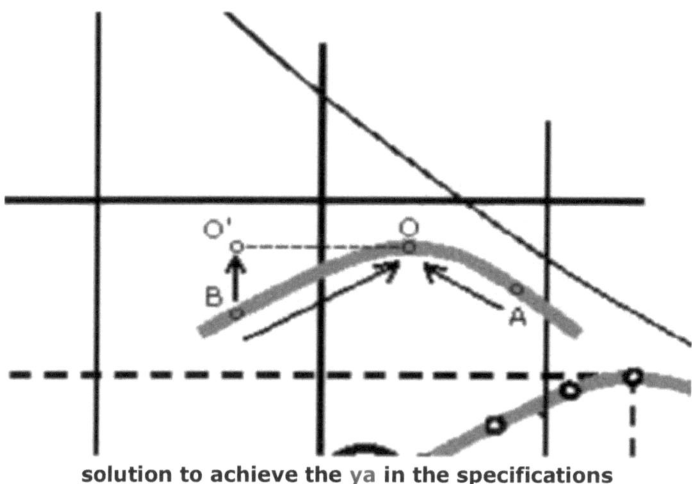

solution to achieve the γa in the specifications

- Once the water content of the soil and therefore the compaction energy required to achieve the required γd are known, the compaction equipment needs to be calibrated, i.e. for a road roller, the minimum number of passes required to achieve the optimum γd over a thickness of around 30 cm needs to be established. The site manager establishes a calibration curve by means of a full-scale test carried out on a section of land. The calibration curve links the number of passes needed to achieve the required , depending on the water content (more or less controlled by watering). This curve passes through a minimum which corresponds to the water content that must be obtained before compaction.

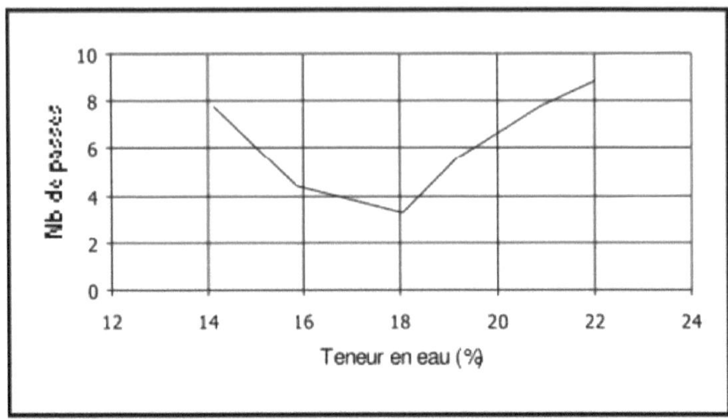

courbe d'étalonnage d'un engin de compactage.

calibration curve for a compaction machine

3.7 In situ compaction verification :

- The problem comes down to taking a ë sample, measuring its volume and weight ($\rightarrow \gamma_h$), and then determining its weight after passing through the oven, which makes it possible to determine γ_d and also w.
- The greatest difficultë resides in the dëtermination of the volume of Sample prëlevë.

Several methods can be used:

- mëthode of the volumëtric ring: a cylinder with a volume of half a litre is gradually pushed into the soil, removing the perimeter as it is pushed in. However, there is a risk of compressing the sample during the operation, resulting in a compaction error. In addition, if the soil contains gravel or pebbles, sinking becomes difficult, if not impossible.
- mëthode dite a l'huile : on exëcute dans le terrain un prélèvement de six à sept litres et on remplissage le trou d'huile, de dens^ connue, à partir d'un récipient qu'on pesée avant et après le remplissage ^ on en dëduit le volume.
- më sand method: same principle, with the excavation filled with dry sand of the right calibre.

- nuc^odensimeter method: this device enables a density to be determined directly by absorbing neutrons emitted by a radioactive source, making it possible to carry out many non-destructive tests on site.
- mëthode of congëlation : saturated sands do not change density when frozen. It is therefore easy to take a ë sample and measure its volume.

3.8 Compaction techniques and equipment :

- The equipment used will depend on the nature of the terrain and the scale of the work to be carried out.
- We use :

> Smooth rollers: these can be "static", i.e. act by pressure alone (these rollers are used less and less, except for rolling surface dressings and compacting very thin mixes < 4 cm) or vibrating, in which case the compaction method combines vibration with pressure.

> Padfoot rollers (also known as "sheepsfoot" rollers): these can be static or vibrating; in addition to the pressure and possibly the vibration, they produce a pëtrissage of the soil and are particularly suitable for compacting clay soils. Smoothing is not necessary after their use.

> Tyred rollers: used for non-cohesive terrain.

> Vibrating plates.

> Pilling engines.

- It should also be noted that whatever equipment is used, compaction must be carried out in thin layers of no more than 20 to 30 cm.
- There are many other techniques for compacting soil.

Application:

The following results were measured during a Proctor test using a normal mould of 0.96 dm^3 . The mass of the mould is 1034g.

Mass of a ë sample of this soil (g)	6,65	6,12	5,02	5,18	5,20	4,77	4,74
Dry mass of sample (g)	6,03	5,51	4,49	4,60	4,59	4,18	4,12
Mass of dry soil with compaction mould (g)	2821	2864	2904	2906	2895	2874	2834

1) Plot the Proctor curve and dëduct the maximum dens^ and optimum moisture content.

2) Calculate the water content at 100% saturation, at maximum density, if G_S =2.67.

3) Dë reduce the amount of water to be added to the optimum to reach saturation.

4) Draw the saturation line at 100% and the line at 5% air.

Assuming relative compaction of 96% at optimum moisture content,

5) Estimate the total stress under an embankment of this soil 20m high.

6) What are the void indices and the degree of saturation of this soil?

Solution:

1) Maximum density and optimum moisture content

Mass of a ë sample of this soil (g)	6,65	6,12	5,02	5,18	5,20	4,77	4,74
Dry mass of sample (g)	6,03	5,51	4,49	4,60	4,59	4,18	4,12
Mass of water (g)	0,62	0,61	0,53	0,58	0,61	0,59	0,62
Water content $\omega = \frac{M_W}{M_S}$							
Water content w%	10,28	11,07	11,80	12,60	13,29	14,11	15,05

Mass of soil with compaction mould (g)	2821	2864	2904	2906	2895	2874	2834
Mould tare = 1034g; Mould volume = 0.96 dm^3							
Dry soil mass $(w)_s$	1787	1830	1870	1872	1861	1840	1800

Dry specific weight $\gamma_d = \frac{W_s}{V}$							
Dry specific weight y^ (kN/m)³	18,61	19,06	19,48	19,50	19,38	19,16	18,75

The Proctor curve is plotted $\gamma_d = f(w)$

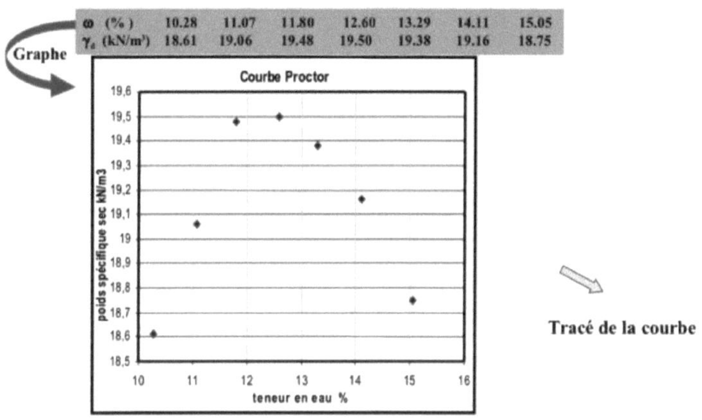

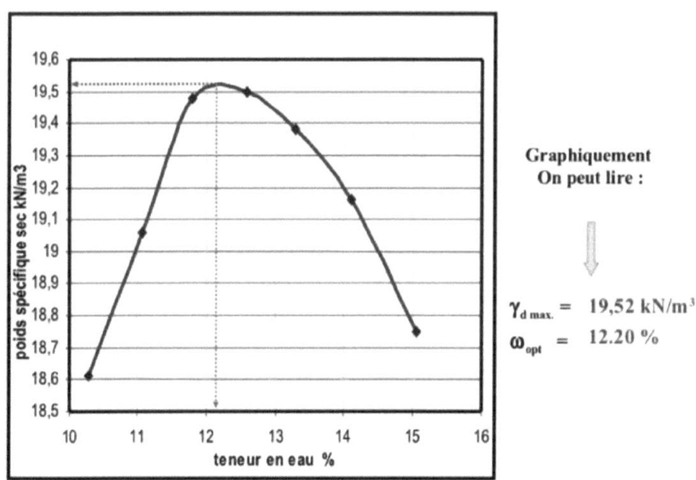

2) Water content at saturation corresponding to maximum density :

$$\gamma_d = \frac{\gamma_S}{\left(1 + w\frac{G}{S_r}\right)}$$

Avec $\gamma_d = \gamma_{dmax}$=19.52kN/m³
γ_s =26,7kN/m³
G=2,67
S_r=1

$$\omega_{sat} = 13{,}77\ \%$$

3) Quantity of water to be added to the P.O. to reach saturation :

$V_\omega = W_\omega / \gamma_\omega$

Comme : $W_\omega = W_s \cdot \omega$

avec : $\omega = \omega_{sat} - \omega_{opt}$

$\Downarrow$

$W_\omega = W_s \cdot (\omega_{sat} - \omega_{opt})$

$\Downarrow$

Par unité de volume (1 m³) on aura :

$\Downarrow$

W_ω = 26,7. (13,77 - 12,20) / 100
= 0,419 kN par m³ de sol

$\Longrightarrow$ $V_\omega \cong$ 42 litres par m³ de sol

4) 100% saturation line :

$$\gamma_d = \frac{\gamma_S}{\left(1 + w\frac{G}{S_r}\right)}$$

With

γ =26.7kN/m3

G=2,67

Sr=1

W(%)	10,28	11,07	11,80	12,60	13,29	14,11	15,05
$\gamma d\ a$ $\underline{Yd}$	20,94	20,61	20,30	19,98	19,70	19,39	19,04

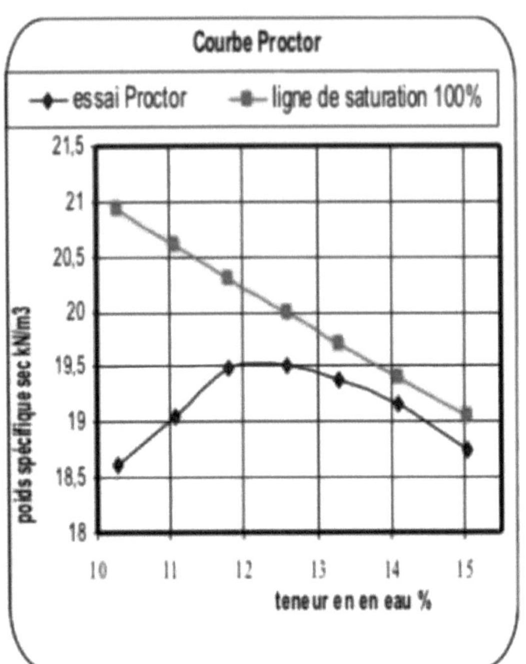

4) 5% air line :

$$\gamma_d = \frac{\gamma_S(1-A)}{(1+wG)}$$

With

$\gamma = 26.7$kN/m3

G=2,67

A=0,05

$w(\%)$	10,28	11,07	11,80	12,60	13,29	14,11	15,05
γ_d	20,94	20,61	20,30	19,98	19,70	19,39	19,04

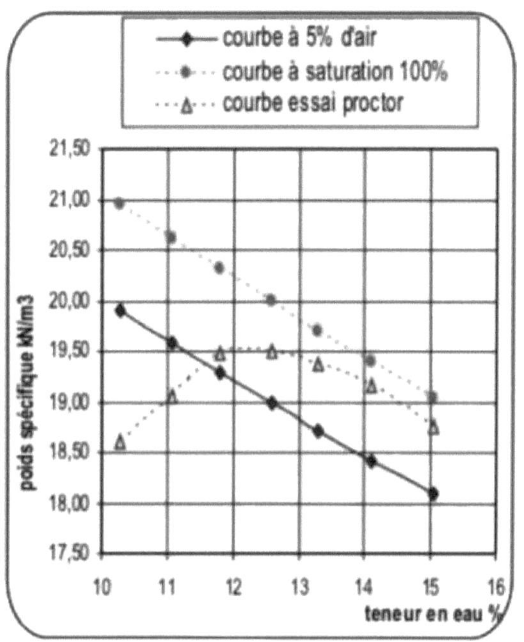

5) Estimation of the total stress under an embankment of this soil 20m high: Relative compaction $\gamma_d * \gamma_{dmax} = 96\% * 19{,}52 => \gamma_d = 18{,}74 kN/m^3$

Total stress : $\sigma = \gamma * H$

Comme $\gamma_d = \frac{\gamma}{1+w} => \gamma = \gamma_d(1+w)$

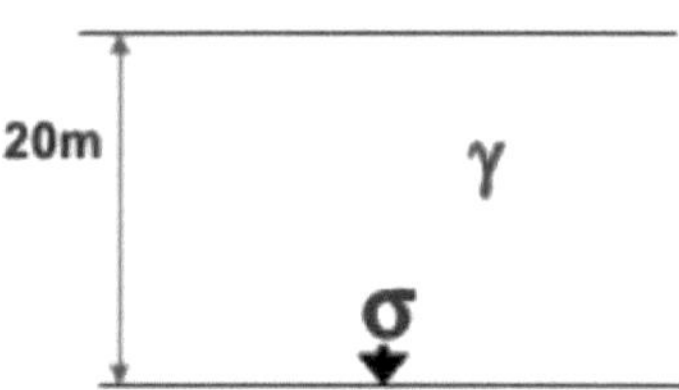

Total density $\gamma = 18.74 * (1 + 12.20\%) = 21.03 kN/m3$ $\sigma = \gamma * = 21.03 * 20 =$ $420.6 kN/m3$

7) Void index and degree of saturation of this soil :

$$e = \gamma_s \left(\frac{1}{\gamma_d} - \frac{1}{\gamma_s}\right) = 26{,}7 * \left(\frac{1}{18{,}74} - \frac{1}{26{,}7}\right) => e = 0{,}425$$

Like $G * w = S_r * e => S_r = 2{,}67 * 12{,}20\% \frac{1}{0{,}425} => S_r = 76{,}64\%$

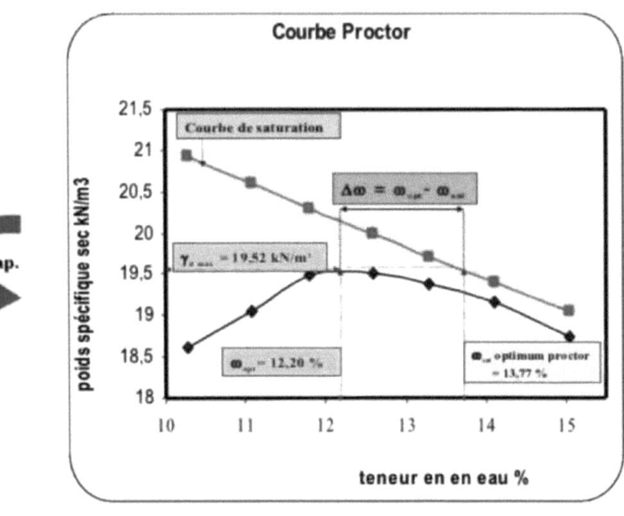

References

[1] Berga Abdelmadjid " Etements de mëcanique des sols " Centre universitaire de Вёск<ar, 2003-2004.

[2] Jacques Lërau " Gëotechnique1 Cours Chapitrel " INSA Toulouse, 2005-2006.

[3] J.Costet G. Sanglerat "Cours pratique de mëcanique des sols, P^з^c^ё et calcul des tassements" Dunod 1977.

[4] Francois Schlosser "Exercices de mëcanique des sols" Presses de 1'ëco1e nationale des ponts et chaussëes, 1989.

[5] Portail de Gënie Civil en A^rie " Mëcanique des sols " www.GCAlgerie.com.

[6] Khaled Meftah " Cours et exercices de mëcanique des sols " First version September 2008.

[7] Fred Portet and Olivier Noel "Gëotechnique routiere, 3- la classification des sols" ENTE d'Aix en Provence, 18 July 2011.

[8] Khelifa Harichane " Mëcanique des solsl " Universe Hassiba Ben-Bouali de Chlef, 2013-2014.

[9] G. N. Smith and Ian G. N. Smith "Elements of Soil Mechanics" University Edinburgh, Seventh edition.

[10] Guettouche Amar " Mëcanique des sols I Cours et exercices " Universe Setif-1, 2015-2016.

[11] P. Vassart " Mëcanique des sols " Bac Construc/IEPSCF - Namur.

I want morebooks!

Buy your books fast and straightforward online - at one of world's fastest growing online book stores! Environmentally sound due to Print-on-Demand technologies.

Buy your books online at
www.morebooks.shop

Kaufen Sie Ihre Bücher schnell und unkompliziert online – auf einer der am schnellsten wachsenden Buchhandelsplattformen weltweit! Dank Print-On-Demand umwelt- und ressourcenschonend produzi ert.

Bücher schneller online kaufen
www.morebooks.shop

info@omniscriptum.com
www.omniscriptum.com

Printed by Books on Demand GmbH, Norderstedt / Germany